FLY SMARTER & SAFER

TIPS FROM MILITARY PILOTS TO HELP YOU SUCCEED IN THE SKY

BY SARAH FRITTS

THINK Aviation

ISBN: 978-1-7321224-0-6

DISCLAIMER

While all attempts have been made to verify the information provided in this publication, neither the author nor the publisher assumes any responsibility for errors, omissions, or contrary interpretations of the subject matter herein.

This book is for entertainment purposes only. The views expressed are those of the author alone, and should not be taken as expert instruction or commands. The reader is responsible for his or her own actions.

Adherence to all applicable laws and regulations, including international, federal, state, and local governing professional licensing, business practices, advertising, and all other aspects of doing business in the US, Canada, or any other jurisdiction is the sole responsibility of the purchaser or reader. Neither the author nor the publisher assumes any responsibility or liability whatsoever on the behalf of the purchaser or reader of these materials.

Any perceived slight of any individual or organization is purely unintentional.

CONTENTS

INTRODUCTION
WHAT THIS BOOK WILL DO FOR YOU

There are a million different techniques pilots use. Which ones are the best? Which ones work?

Well, why not use the proven techniques of military aviators?

The military has been involved in aviation since the very first flight.

They know their business. If you take what they do and apply it to your flying, you will become a better pilot.

This book will show you what our nation's finest aviators do on a daily basis and give you practical tips to help implement those practices into your own flying.

These military tips will take you from average pilot to awesome pilot.

Or, at the very least, a safer one!

MILITARY TIP #1
PLAY BINGO

The military is obsessed with fuel numbers and on-station time. We need to know down to the minute how long we can stay in the air supporting the troops on the ground. So we created a term called "bingo fuel."

"Bingo fuel" is the minimum amount of fuel you need in your tanks to get back home. This number is NONNEGOTIABLE!

It means you MUST head home, without delay. It doesn't mean you are in an emergency situation, it just means you gotta get home...now.

I will admit, this number doesn't have a ton of application in general aviation, as it does in the military, but there are situations where you would want to use it.

As a United States Army aviator, most of my missions begin and end at the same location. I fly out to a point and loiter for hours while working with a ground unit. In this instance my bingo fuel number is super important. Once I get on station I know exactly how much time I can give the ground commander before I have to head back to base.

But how does that relate to you? Well, what if your best friend is in town and you want to fly out to Mount St. Helens, loiter for a while, head over to Mount Adams, hit Mood Hood and come back?

Your bingo fuel, in this case, alleviates the need to constantly calculate fuel numbers and/or stick to a rigid flight plan schedule. Your bingo fuel number lets you fly around until you hit that number and then you head home. Want to stay at one mountain longer than the other? No problem.

To calculate this number, simply pick the fuel required to fly to the farthest distance from the airport you will fly that day, add your VFR fuel reserve requirements and that is your "bingo fuel."

For example, let's say Mount Adams is the farthest away you will fly that day, and it's about an hour away from your airport. If your fuel burn is about 10 gallons an hour and you need a 20 minute reserve (I like to use 30 minutes instead), you need to head home when you have 15 gallons left in the tank.

15 gallons is your "bingo fuel" number.

Remember, it's nonnegotiable. Don't let your passengers try and convince you to stay a little longer. It's a hard number.

The only caveat I will say is make sure you compare your actual burn rate to your planned burn rate. If it's more, then you will need to add a little to your bingo fuel number.

The bingo number doesn't relieve you from doing a full flight plan or detailed fuel planning. It raises your fuel awareness, gives you flexibility and gives you a margin of safety. I hope that helps. Try it sometime!

By the way, "Bingo fuel" is a military term, so avoid using it with Air Traffic Control.

MILITARY TIP #2
CONDUCT AFTER ACTION REVIEWS

An instructor of mine said something I will never forget:

"The hours in your logbook only count if you were paying attention."

One of the best ways to pay attention, besides staying awake, is to take notes of the things you messed up during the flight. When you are done with the flight, use those notes to conduct an After Action Review (AAR).

I believe one of the reasons the United States Army is the best in the world is our obsession with After Action Reviews.

Every training and combat event in the Army ends with an AAR. It's so important, it's in our preflight mission brief checklist. It says: "Decide on location for AAR after flight." Before we even take off, we pick a location where we will discuss what happened on the flight.

An AAR can be as complex or simple as you need it to be. At the

National Training Centers, AARs can take a day. Your AAR at the side of the plane need not take that long!

AARs only need to contain a few simple elements:

- 3 things that went well
- 3 things you could have done better

I also like to add in these things:

- Discuss any crew coordination issues you and your copilot had. For example, did they touch a switch without announcing they moved the switch? Did they forget to do a before-landing checklist even though you asked them to? You get the picture.
- Research items. Sometimes an event during the flight will make it apparent you have a gap in your aviation knowledge. It could be a regulation, aircraft system or aircraft limitation. Ideally, you should look into it as soon as possible, but at the very least make a note and research it before your next flight. I find I remember regulations much better when I can tie it to an experience.

Don't underestimate the value of capturing and correcting your lack of knowledge. Take it from Chuck Yeager, the famous record-setting Air Force pilot:

"I was always afraid of dying. Always. It was my fear that made me learn everything I could about my airplane and my emergency equipment, and kept me flying respectful of my machine and always alert in the cockpit."

Conducting an AAR can be the most difficult part of your flight. Remember when your instructor constantly pointed out your shortfalls? The constant barrage of "you suck" was exhausting, wasn't it?

When formal training ends, you get to run from that feeling.

But, stop! You're an aviator!

Turn around and run back. Lean into that horrible feeling of worthlessness and embrace it. Use it to get better.

Yeah, I know it sucks coming to terms with your shortfalls. It takes great courage and maturity to admit you messed up. But aviation requires you to be brutal on yourself.

Start doing AARs. They are vital for your development as an aviator.

To steal another quote from the great Chuck Yeager: "There's no such thing as a natural-born pilot."

It takes work.

MILITARY TIP #3

HIT THE BOOKS

You need to know your airplane's operator's manual inside and out.

In the Army we call it the "Dash 10" because of how the training manuals are labeled:

TM 1-1510-225-10

The military treats the aircraft's manual like gold. It's the starting point for all of our training.

You should also know your operator's manual inside and out.

Specifically learn the limitations, emergency procedures and normal procedures chapters. These chapters are mandatory.

You must be able to regurgitate them at any time.

Army aircraft manuals are laid out the same. Chapter 5 contains limitations, chapter 8 contains start-up/preflight procedures and chapter 9 contains emergency procedures.

CHAPTER 5	OPERATING LIMITS AND RESTRICTIONS ......
CHAPTER 8	NORMAL PROCEDURES R ....
CHAPTER 9	EMERGENCY PROCEDURES...

Ever since I started flight school, knowing these chapters was essential. I had to know the emergency procedures and limits word for word, no exceptions..

In the civilian world those same chapters are chapters 2, 3 and 4. Check out this Cessna pilot's operating handbook (POH):

TABLE OF CONTENTS

	SECTION
GENERAL	1
LIMITATIONS	2
EMERGENCY PROCEDURES	3
NORMAL PROCEDURES	4
PERFORMANCE	5

I highly suggest you pick up your aircraft's POH right now and skim through those chapters. Is there anything you don't know by heart in those chapters? Why not? This information is paramount to your safety.

You should also go through your operator's manual and learn all the NOTEs, CAUTIONs and WARNINGS. They contain an incredible amount of useful information that will keep you from killing yourself and/or damaging your aircraft.

Check out a note from the Cessna POH:

NOTE

Takeoff and land with the fuel selector valve handle in the BOTH position.

Like I said, there are some really good extra bits of information in the NOTEs, CAUTIONs and WARNINGs.

So, do yourself a favor and act like Army pilots! Memorize chapters 2, 3 and 4 of your operator's manual. Then move on to the NOTES, CAUTIONs and WARNINGs.

Everything else is gravy.

MILITARY TIP #4

DO A 10-YARD DASH

In the last tip I talked about knowing your operator's manual. In this tip, I want to go a step further: take an open-book operator's manual test every year.

Every year Army pilots have to complete a myriad of tests, classes and check rides. One of those requirements is to sit down with the operator's manual/handbook and take an open-book "test." We have to methodically go through the manual and answer questions. We call it a "Dash 10 test."

The intent of the operator's exam is to get you deep into the nitty-gritty of the manual. The Dash 10 test is also an excellent refresher on topics you though you would never forget when you were in training.

Every operator's manual contains an incredible amount of useful information. But the only way to remember it is to go through the manual every year.

As a civilian GA pilot, you will have to create one for yourself. Or you could ask a fellow pilot to create one for you.

If you are an instructor pilot, I highly recommend you create one for your students.

Here are three guidelines to follow when creating an exam:

- Make the questions worth asking. The answers should be important and relevant to operating the aircraft.
- Make the questions in order. Question 1 should be from the front of the manual, and question 50 should be toward the end. This aids in the methodical movement through the manual.
- Broaden the Dash 10 test to include any extra equipment and avionics you typically fly with (e.g., include some questions from the Garmin 430 manual). Also add any FAA directives specific to your aircraft.

So, I know most of you have never seen a Dash 10 test, which is why I am linking to two I created.

You can find both Dash 10 tests on the Fly Safer and Smarter resource page[1].

1 thinkaviation.net/fly-safer-and-smarter-resources

MILITARY TIP #5
GET CHECKED OUT

The last couple of tips cover how the military requires its aviators know their operator's manuals inside and out. But thats not all.

The military also understands the importance of continual practice. They do that by requiring a yearly check ride.

You too should take a yearly check ride. Yes, I know. You aren't required to have a check ride with an instructor every year. So?

"Check rides," if you find the right instructor, are a phenomenal opportunity to learn something new and practice rusty emergency procedures. Studies have shown nine months is the amount of time it takes to lose proficiency on emergency procedures.

Yes, it's expensive to pay for an instructor. So what? It's not as expensive as plowing your plane into the ground doing something utterly preventable.

Flying a plane is not like riding a bike. At all.

Emergency procedures don't just "come back to you" in high-stress situations unless you practice them. Check rides encourage regular practice. I was able to avoid crashing a King Air that yawed on me 10 feet over the ground because a few days earlier I had practiced go-arounds.

Practicing works.

What should the check ride cover?

The check ride should consist of two or three things:

1. Oral exam: topics include emergency procedures, limitations of the aircraft, normal procedures, airspace, regulations, aircraft systems, aerodynamics, weather and so on.
2. Standardization flight exam: demonstrate all normal and emergency procedures.
3. Instrument flight (if you are rated): to include hand flying an approach. If you are multiengine rated, then hand-fly a single engine ILS approach.

Please resist the temptation to find an easy instructor. Find one who will ask you the hard questions and help you narrow in on your knowledge gaps.

BONUS: Write down every question you missed. Go back and study these topics throughout the year, focusing specifically on your weaknesses.

If you have never done aerobatics before, then find an instructor capable of doing spins. Have them show you full spins until you can do one without peeing yourself.

Do you see a theme in the professional aviation world?

Regular practice, regular studying and recurrent training.

You too can act like a pro.

MILITARY TIP #6
DO A "SAFETY STAND DOWN DAY"

Flying is a dangerous business.

The Army knows this and has put four controls in place to reduce accidents. I want to talk about #4 on the list.

1. We fill out risk-assessment forms before every flight. The pilot in command (PIC) must brief that form and get signatures from another PIC and the commander before flying.
2. Every unit has a trained safety officer assigned.
3. We have forms we fill out when we see something is unsafe, which get addressed at the monthly safety meeting.
4. We also conduct a yearly "safety stand down day."

Sometimes, when things get really bad, like they did a couple of months ago when we lost several aircraft and lives, we do impromptu safety stand down days.

mountain or a hill if you can't see what's on the other side.) I didn't find him, which I'm thankful for, as he apparently exited the front windshield on impact. Cause of death: massive head trauma.

Why am I telling you this?

If he had used a spot tracker, we would have found him quickly. He told his family where he was going, but he didn't have real-time flight following. Radar coverage ended 30 minutes before he crashed, so flight following didn't help him much. His family had to wait in agony while we searched for days.

Do you see why you need to think seriously about flight following?

Now that I've got you thinking, here are some options for real-time flight following:

1. Use a Spot Tracker[4]. I love, love, love these things! If you are a hunter, a hiker, a mountain climber, a fisher or an airplane pilot, go get one! It interacts with satellites for real-time tracking.

You can send a distress signal as well. Brilliant.

I used one of these when I flew all over Oregon one summer. I knew no one would ever find my green helicopter under a 150-foot doug fir canopy. The Spot Tracker gave me the peace of mind that rescuers could find me quickly if something went wrong.

4 amzn.to/2hAgQ7C

You can read more about Personal Locator Beacons and Satellite Messengers on REI's website[5].

Yeah, they cost money, but they are worth every penny.

2. File an IFR flight plan if you are instrument rated. This counts as real-time tracking. Search and Rescue will get activated quickly if they can't contact you.
3. As a VFR pilot, request flight following. This is completely different from opening a flight plan with FSS.

Flight following is a must, especially at night and/or over sparsely populated terrain.

Need a refresher on picking up flight following? Check out this article I wrote: "*How to Pick Up VFR Flight Following*[6]."

Just keep in mind, flight following can't help you when there is no radar coverage. I know a lot of VFR pilots abhor talking on the radio and can't stand getting flight following. If that's you, then you need to pony up some money and get a Personal Locator Beacon or Spot Tracker.

It could save your life or save your family days of heartache.

---

5 www.rei.com/learn/expert-advice/personal-locator-beacons.html

6 thinkaviation.net/how-to-pick-up-vfr-flight-following

MILITARY TIP #8
NEVER FLY SOLO

Don't fly single-pilot IFR.

In the Army, our regulations prohibit single-pilot IFR flights because it's dangerous.

It's even more dangerous without an autopilot.

Most IFR accidents occur during the takeoff, climbout, approach and landing phases. But those portions of the flight are usually less than 25 percent of the total flight. A second pilot is invaluable during these phases of flight.

I'm not saying the other person should be IFR rated, although that certainly helps, but they should be a pilot. You need someone who can tune radios and NAVAIDs, and perform other tasks without supervision when directed by you.

I have had many a flight where everything seemed fine, but once I descended into the clouds it was like I had entered a whole new world. Add in turbulence (which inevitably happens when you enter the clouds) and poor Air Traffic Control vectors and suddenly a beautiful flight turns extremely stressful.

A second pilot is invaluable in these situations. You can fly the aircraft and they can focus on the radios. I can't tell you how many times I was glad to have someone sitting next to me during the approach phase.

Everything changes when you're in the clouds.

MILITARY TIP #9
COMPRESS YOUR TRAINING

I want to talk about the importance of compressing your training.

The military understands this concept really well. Training five days a week produces more proficient pilots. You get good fast.

I understand you may not have the luxury of training every day.

But wait...are you sure about that? Let's challenge that assumption.

I hear two main arguments against training every day:

1. I don't have the money.
2. I don't have the time.

Let's address both:

I don't have the money:

This is an expensive hobby. I am all for general aviation being accessible, but let's face it, aviation costs money.

Most insurance companies don't consider you proficient unless you fly five hours a month after you have your rating.

Think about that for a second. Once you get your license, are you prepared to fly twice a month for at least two hours each? That's two weekends or three weekdays a month. Can you do that?

More important: Can you afford that?

You need more money up front when you go through training. If you only have the money to train twice a month, I am going to be brutally honest with you: wait until you have the money to fly at least twice a week during training.

Otherwise you will constantly have to start over. It will take you way more time to get through the training, and it will cost you more in the long run.

For those of you trying to get an aviation job, you need to take out a loan and get your training done ASAP. You absolutely cannot nickle-and-dime your aviation training if you are pursuing an aviation career.

I don't have the time:

We don't have enough of it, right? We all feel that way. But you have options.

I sit down at my desk at 5:15 a.m. every morning to write for my blog because it's important to me, and it's the only time I have.

I don't say that to brag, but rather to emphasize that when something is important, you make time for it.

When people tell me what's important to them, I take it with a grain of salt, because talk is cheap.

Where you spend your time is a direct reflection of your priorities.

If you aren't willing to get up early to study and practice your craft, is it that important to you? Are you sure you want that aviation rating? Because if you are getting a rating, you should be studying or flying five days a week until it's done.

Once it's done you need to keep flying several times a month and study weekly.

So let's talk solutions to these two problems.

When you're short on time and money, you need to find some creative ways to condense your training. Here are some examples:

1. Get a home simulator program and fly it on the days you aren't in the actual airplane.
2. Use the Quizlet app[7] and study flash cards in the morning. I use it almost every day. It's awesome.
3. Listen to live ATC feeds[8] (especially if you are getting your instrument rating).
4. Take out a loan.
5. Get your family, coworkers, boss and friends on board with

---

7 quizlet.com/mobile

8 www.liveatc.net/feedindex.php

your dream. Tell them you will be busy two to three days a week until you get your rating.

This stuff is perishable. It's perishable for the professionals, and it's even worse for hobbyists. You need to get your training done as soon as possible.

It will make you a better aviator and save you money in the long run.

MILITARY TIP #10
JUST SAY NO...TO INEXPERIENCE

General aviation is a backward industry. It's downright bizarre.

Tell me: In what profession do the most inexperienced people instruct?

In most professions or trades, you have to have some semblance of experience to teach. Not so with general aviation. A brand-new aviator with barely 250 hours can instruct.

This never happens in the military. The idea of a 300-hour pilot instructing anyone is ludicrous.

You are lucky if you can act as a pilot in command ("captain" in civilian speak) if you have 350 hours. Then you need at least 50 hours as PIC before you can apply for a specialty like instructor pilot, maintenance or safety.

So, at a minimum you need about 400 hours before the military will consider you for instructor school, and that's really low. As a commander, I wouldn't send anyone unless they had 600 hours... minimum.

The military doesn't let brand-new aviators teach, and you shouldn't accept that standard for yourself either.

Find yourself an experienced instructor.

I know, they are so hard to find! I have read countless emails from my newsletter subscribers about how their instructor left and they couldn't find another at the local Fixed Base Operator (FBO). It's a real problem.

If you are at a Part 141 factory schoolhouse, you are more likely to run into the churning instructor pool. You may need to go to a smaller airport or the FBO and hunt down one of the older instructors.

Getting flight instruction from an experienced pilot will do a few things for you.

1. You will avoid the turnstile of young instructors going to the airlines.
2. You will get your questions answered by someone who has *actual* experience.
3. You will most likely get a higher quality of instruction.

It may cost you a bit more, but I think it's worth it.

If I ran a flight school, I would never ever hire anyone with fewer than a 1,000 hours of real-world, non-traffic pattern hours.

But that's just me. I grew up in a world where instructors are the most experienced pilots in the unit. It was a coveted role and a respected one.

MILITARY TIP #11
CHECK YOUR CHECKLIST

No one is perfect. No one can remember everything all the time.

This is especially true in the aviation world. The military has known this for a very long time. Compared to other professions (like surgery), we have been using checklists forever.

The military relies heavily on checklists for two reasons. We fly with two-person crews, so it's not necessary to memorize the checklist. The other reason is the equipment we fly is complicated. It's hard to remember everything.

In general aviation you are more likely to fly simple aircraft as a single pilot. However, that doesn't mean you shouldn't use checklists.

The military and airlines could mandate their pilots memorize checklists, but they purposefully do not. (Note: They both do to some degree by using "flows," but "flows" are usually backed up by abbreviated checklists.)

There is a danger to memorizing checklists. During high-stress environments and ATC interruptions, it's extremely easy to forget a vital step. Even with checklists, professional pilots forget things all the time!

So, I propose a middle ground:

Go ahead and memorize the checklist, but skim through the checklist after you have completed the steps.

You'll be amazed how many times you forget things, especially if you get interrupted by ATC.

Also, consider always using the checklist on more complicated areas of flight like the "Start" and "Run-up." "After-takeoff" and "Cruise" checks aren't usually as intensive and lend themselves to memorization with a double-check.

Here are a few other checklist tips:

1. Use your finger to follow along so when you get interrupted, it's easy to find your place.
2. If you get interrupted once, continue with the checklist, but if you get interrupted twice, start the whole section over again. (I know this may seem nit picky, but a lot of professionals use this technique.)
3. If you are an infrequent flier, then perhaps you shouldn't bother memorizing the checklist. You are more apt to forget something compared to daily fliers.

4. If you are a two-pilot crew, always start the checklist by naming the section (e.g., “Run-up”), and then go to step one. Don’t just start with step one without announcing which section you are on. Also, when you have finished the section, announce “Run-up check complete” to let the other pilot know you have finished the section.

Oh, and one more thing: go ahead and preflight with no checklist so you don’t leave it on an engine compartment...but make it a habit of reading the checklist carefully after you are done. When you miss something, which you will, go back and check the things you missed.

MILITARY TIP #12

TAKE A HIKE

This tip is one I hope you've heard before. If not, you need to add it to your bag of aviation habits immediately:

—Do a walk-around before you get in the aircraft—

Finish your paperwork, do your preflight, hit the bathroom and then right before you get in the aircraft, walk around.

Make the walk leisurely. This is your last chance to catch obvious stuff like tie-downs.

Here are some techniques I use:

1. I walk fairly close to the aircraft. I do this so if the aircraft is still tied down, I will trip on the tie-downs. Don't laugh: people start aircraft all the time with the aircraft still tied down or covers still attached.
2. Pick up the chocks as you pass the nose (or side) wheel.
3. Study all the fasteners to make sure they are secure.

4. If you can reach them, touch the fuel caps and make sure they are secure. If you can't touch them, then look at them closely.

That's it. Pretty simple, but it's a powerful tool.

I also like to take it a step further:

I study the aircraft as I walk up to it for the first time before I start my preflight. Viewing the aircraft from a distance can reveal some things you may not catch up close.

Plus, studying the aircraft puts you in a "flight" mind-set. Deliberately focusing on the aircraft will put you in business mode.

You may also want to consider putting the phone away until you have taken care of the preflight. I notice the phone is extremely distracting. I recently started severely restricting its use from preflight to the end of the flight. If you are in training or taking a check ride, I highly recommend it stays in the car unless you need it for flight planning and weather checking.

I won't lie, I have sent many a text/email before a flight, but I don't do it anymore unless it's related to the flight. I don't like how checking my phone takes me out of my "flight" mind-set. Aviation is a serious business. Treat it as such.

Take it a step further: Do a walk-around after your flight. Make this nonnegotiable.

Look for damage from bird strikes and any other issues, and always check the oil and props.

This is especially important after night flights. Many morning flights have been canceled because the night pilots were in a hurry to leave and didn't find a problem someone could have fixed overnight.

So, if you aren't doing this already, do a walk-around before and after the flight. Oh yeah, and I don't care if you fly with another pilot with 10,000 hours or if you are in training.

You should always do your own walk-arounds.

MILITARY TIP #13
DON'T LEAVE HOME WITHOUT IT

Almost every military aviator flies with some type of vest. Have you ever wondered what's in these vests?

The vests contain everything a pilot needs to stay alive while they run from the enemy.

Here's a list of things on my vest:

- Radio (more about this in the next tip)
- One-handed tourniquet (it's in the same place on everyone's vest)
- Flashlight (I carry extra batteries)
- Big badass military knife
- Basic medical supplies
- Signaling mirror
- Carabiners and rope for extraction

- Pouches for weapons magazines (I keep my sunglasses and a protein bar in them while in the United States)

- One-handed seat belt cutting device

There are a few more things, but you get the idea.

This is quite a long list, but let me boil it down to the four most important (in my opinion):

1. The emergency radio

2. Tourniquet

3. One-handed seat belt cutter or substantial knife

4. Flashlight (with extra batteries)

Why those things? Let's go through them.

The emergency radio: Communication is the most important element to a successful rescue. Always fly with a hand held emergency radio. Check the batteries during your preflight.

Tourniquet: Of all the injuries you could get, excessive bleeding is what will kill you in minutes. Head trauma, broken bones and other "major" injures are usually manageable for a few hours. Not so with arterial bleeding.

I strongly believe in the importance of tourniquets. I never go hiking without one. It is the one piece of equipment you need in the middle of nowhere that will save someone's life. I highly recommend a one-handed tourniquet. The Army understands this concept so well, they require every soldier in Afghanistan and Iraq to carry one in their lower leg pocket.

Go here for more information on military tourniquets:

amzn.to/2nrTvYk

Flashlight: A flashlight is the third most important element and let me explain why:

When we conducted a search-and-rescue training exercise, the searchers on the ground tried out different types of signaling devices while we flew around. The flashlight outshone all other devices, even the fancy flares. Oh yeah, we did this experiment during the day.

Never ever fly without a flashlight. You can also use it to check the wings for icing.

Knife or one-handed seat belt cutting knife: Just remember to keep a couple within reach. I heard a story of an Army helicopter pilot who'd crashed and then dropped three knives trying to cut himself out because he was shaken up so bad. Luckily, there were people around to extricate him (and watch him drops his knives).

I know GA pilots don't wear vests, which is a bit of a shame really. While you don't need to run away from the enemy, you may become trapped in the cockpit and will want those supplies within reach.

Keep this in mind: A radio, tourniquet, knife, flashlight and water/food are useless in the cargo compartment.

You should also seriously consider packing a "go" bag (a term we use for if we had to grab our bag and run from the crash site). It should have additional supplies in case of a precautionary landing.

We also fly with a survival bag. It has enough food, water, cold-weather gear and additional medical supplies for the entire crew to last three days.

One thing I didn't cover is the fire extinguisher and first aid kit. Every aircraft should have its own first aid kit and fire extinguisher. Both should get inspected periodically.

If the planes you rent don't have first aid kits and fire extinguishers, bring your own.

BONUS: Never fly without water and a jacket no matter how hot it is, mundane or short the flight!!

Even if it's warm out, if you have to spend the night somewhere, you will want water and something warm in the wee hours of the morning.

MILITARY TIP #14
ROGER, ROGER

I want to highlight one specific element of the ALSE equipment that is more important that the rest:

An emergency radio

The survival radio is a nonnegotiable piece of equipment on Army aircraft.

No radio = No flight

You too should consider an emergency radio a requirement.

But, wait! I have a cell phone! Will that work?

Please, please, please do not rely on your cell phone!!! Why?

Because cell phones can't talk to rescue aircraft. With a radio, you can talk to your rescuers and relay vital information.

More important, people forget to charge cell phones, and they have awful battery lives (24-hour max vs. days for a survival radio). They also have terrible coverage just when you need them to have amazing coverage.

You get the point: a cell phone is NOT a survival radio.

I am going to take it a step further and say always carry a personal locator beacon (PLB) even if you use VFR flight following.

I briefly touched on the PLB in military tip #7, "Always Have Flight Following." The PLB will get emergency crews to you faster. They are highly accurate, and your family will know where you are.

I have done many search-and-rescue missions that could have been resolved in hours if the person had a PLB. I wish people who go hunting/hiking/flying in the middle of nowhere would buy a PLB!!! Especially 76-year-old diabetic bow hunters who are impossible to find in the brush because of their camouflage.

I know aviation is expensive. While Foreflight on your iPad is an option, an emergency radio is not an option. It should be your second purchase after your headset.

Go to the Fly Smarter and Safer resource page[9] for suggestions on radios.

BONUS: You should consider always putting your emergency radio within reach when you are sitting at the controls. If you become trapped in a crash, you will want to be able to reach it. It is not useful in a bag in the back seat.

---

9 thinkaviation.net/fly-safer-and-smarter-resources

Army aviators carry their radio in a vest they wear, which also has a tourniquet and a seat belt cutting device/knife. Think carefully about where you want to put your safety equipment.

MILITARY TIP #15

LOOK LIKE TOM CRUISE

Okay, let me explain because we can't look like Tom Cruise, but we can work out like the fighter pilots in *Top Gun.*

This military tip transcends all services:

Exercise five days a week and eat "clean."

Aviation is a strenuous business. It demands mental acuity, alertness and good overall health. Have you ever seen a fat fighter pilot? Me neither. Actually, I hardly ever see fat airline pilots either (major airlines, not regionals).

Speaking of which, I noticed Alaska Airlines does a good job of hiring fit pilots. Most of their captains have a super healthy body weight. I have a feeling they hire this way on purpose.

If you want to continue flying into your late 70s, it takes a lifetime of good habits. You won't have to worry about retaining a medical if you eat clean and exercise consistently.

Here are two resources you can use to correct health issues:

Sleep Smarter[10] by Shawn Stevenson (he also has a podcast called *The Model Health Show*[11])

It Starts with Food[12] by Dallas and Melissa Hartwig

Notice I didn't even recommend an exercise book? Food and proper sleep is 80 percent of the battle.

I'm also not going to recommend the type of exercise you should do. I personally prefer interval/circuit training, but it's not for everyone. It's more important you do what you love for long-term health.

Even walking is great. Have you tried to get 10,000 steps a day for a seven days? It's hard! Or at least I find it hard. I don't think I have ever done a seven-day stretch, actually.

Remember, though, it's food, not exercise that will help you keep your flight physical.

You cannot exercise your way out of a marginal diet.

BONUS: For a completely crass and non-PC book, try *Fitness Confidential*[13] by Vinnie Tortorich. His advice is simple: no grains, no sugar. He also has a podcast.

10 amzn.to/2lDVAD8

11 theshawnstevensonmodel.com/podcasts

12 amzn.to/2yXPzCZ

13 amzn.to/2grJc42

MILITARY TIP #16
DON'T PASS ME A WRENCH

I can't speak for the other services on this one, but in the Army, each warrant officer is expected to "track" an aviation specialty.

There are four "tracks": instructor pilot, maintenance pilot, tactical operations and safety officer.

Warrant officers can't track until they have at least 50 hours as a pilot in command. Our definition of PIC is not the person at the controls. We define PIC as the "captain" or the ultimate authority of the aircraft. It takes experience and maturity to become a PIC. It usually takes around 500 hours to become a PIC, to give you some perspective.

Back to the tracks...I recommend you also pick a track.

I know civilians can't go to the military courses, but there are civilian equivalents:

Instructor Pilot: get every CFI rating available

Maintenance Officer: get your A&P license

Safety Officer: go to OSHA training, sign up for online classes at Embry Riddle University, attend AOPA safety briefings regularly and read safety magazines

There is no civilian equivalent to tactical operations officer. They deal with combat personnel recovery, mission planning and other combat specific equipment.

Which track should you choose? Trust me on this one: you will know which one you want to do.

For example, anytime someone asks me if I want to go on a maintenance test flight, I can find a million other things I would rather do. But I will spend hours explaining VFR chart symbols to a new pilot. Obviously, I would never track maintenance.

Pick something you enjoy and specialize in it.

Your aviation career is just that: a career. I don't care if you only fly once in a while. It's still a career. Aviation is a serious business and you have to treat it like a profession regardless if you're paid to fly or not.

MILITARY TIP #17
RISKY BUSINESS

The military conducts dangerous business, which is why it has created a system to identify and mitigate the risks it takes.

I noticed the civilian world isn't as rigorous when identifying risk, but that doesn't mean you can't conduct your own risk assessment prior to flight.

Risk assessments in the military are not there so the commander can say "no" to a flight. They exist to help commanders identify what they need to change so they can say "yes." There is a huge difference between the two mind-sets.

For example, sometimes our unit conducts recovery missions on Mount Hood. It's easy for one of those missions to get into the moderate- or high-risk categories. However, because we have a method to identify why it's high, we can modify or "mitigate" the risk. Is it high risk because the crew is inexperienced? Is it high risk because of the weather? We can either change the crew or we can delay the flight until the weather gets better to mitigate the risk. Without a risk assessment, there is no way to pinpoint the cause of the risk, though. Sure, you could do it in your head, but if you are inexperienced, it's hard to remember all the hazards that could affect your flight.

I don't have a civilian equivalent of a risk assessment, but I have a military version. We use a different one for combat operations, in case you think I'm giving away state secrets.

Take a quick look at it. You can also download it here:

thinkaviation.net/fly-safer-and-smarter-resources

(1) COMMAND RELATIONSHIP	HIGHEST=
Assigned/Organic	1
Attached/ Direct Support / OPCON	3

(2) MISSION PLANNING	HIGHEST=		
	Planning Time		
	>4 Hours	1-4 Hours	<1 Hour
SPECIFIC	1	2	3
IMPLIED	2	3	4
VAGUE	3	4	5

(3) MISSION COMPLEXITY	TOTAL=			
TASK/MISSION	DAY	NIGHT	NVD	IMC
Admin / Service	1	4	3	2
Continuation/ATM Training	1	3	2	1
Counterdrug	2	4	3	NA
External Loads / Water Buckets	3	NA	2	NA
Ferry Flight	1	3	2	1
Fire Support	3*	5*	NA	NA
Hoist Operations	2	NA	3	NA
Maintenance Recovery (note 2)	2*	4*	3*	NA
MTF / FCF (note 2)	3	NA	4	NA
Multi-ship	1	NA	3	NA
RL Progression / Eval	1	3	2	1
SAR/MEDEVAC	3	5	4	3

(4) ADDITIONAL FACTORS	TOTAL=		
CBRN	3	Live Hoist*	5
CEFS	2	Migratory Birds	1
Confined Area	3	Mountain Ops	4
Dust/Snow/Sand Landing	4	OGE Pwr within 5% max tq	3
Emergency Procedures	2	PC <50 hours in MDS	2
FLIR – EO/IR	2	Terrain Flight	3
IGE Pwr within 5% max tq	5	Unsurveyed LZ	3
IMC Alternate Required	2		
Laser Ops	3	UH-72A	
		LS-16 (Night Sun)	2
		Photo / RECON / Observation	1
		Single Pilot FCF	5
		Single Pilot**	5

(5) CREW SELECT TOTAL TIME			TOTAL=	
PC=	PI=	NCM=	NCM=	NCM=
RCM	≥2000	<2000	<1000	<500
NCM	≥500	<500	<250	<100
LAO ≤50	1	2	3	4
LAO >50	0	1	2	3
<100 in MDS	Add 2 per crewmember			

(6) FIGHTER MANAGEMENT			TOTAL =	
PC=	PI=	NCM=	NCM=	NCM=
			Hours of Rest	
			≥10	<10
Residence			0	1
Garrison (Private room barracks / hotel)			2	4
Field (Open bay barracks / tents)			3	6
Duty Day Total 0-12 Hours				0
Duty Day Total 12-14 Hours				2
Duty Day Total 14-16 Hours				4*
Duty Day Total >16 Hours				8*

(7) WEATHER (CEILING/VISIBILITY)			TOTAL=	
Flight Conditions	≥3000/5	>1000/3	<1000/3	<600/2
DAY	0	2	4	6
NIGHT	2	4	NA	NA
NVG	1	3	5*	8*
IFR	0	0	2	4

ADDITIONAL FACTORS DD 175-1			
ICING	LT = 1	MOD = 2	SEV = NA
PRECIP	MOD = 2	HVY = 3	
TURB	LT = 1	MOD = 2	SEV = NA
SFC WIND	>20KTS = 2	>30KTS = 3	>40KTS = 4
T/STORMS	3		

(8) MOON DATA (max value of 5)	HIGHEST =	
	Moon Angle Above Horizon	
Illumination %	<30°	≥30°
>60%	5	0
30 – 60 %	4	2
<30%	3	3
Broken or Overcast Sky Conditions	4	
No moon	5	

(9) NVG CREW SELECTION			TOTAL=	
PC=	PI=	NCM=	NCM=	NCM=
Last NVG Flight	Total NVG Flight Time			
	≥150	<150	<100	<50
<30 Days	0	1	2	3
≤60 Days	1	2	3	4
Uncurrent	2	3	4	5

RISK ASSESSMENT BLOCKS	TOTAL FROM BLOCKS	
	PER 1/2	PER 2/3
1. COMMAND RELATIONSHIP		
2. MISSION PLANNING		
3. MISSION COMPLEXITY		
4. ADDITIONAL FACTORS		
5. CREW SELECTION		
6. FIGHTER MANAGEMENT		
7. WEATHER		
8. MOON DATA		
9. NVD CREW SELECTION		
TOTAL RISK ASSESSMENT		

TOTAL RISK ASSESSMENT - CREW CONFIG				
# CREW	LOW	MOD	HIGH	EXT HIGH
1	0-15	16-30	31-40	41+
2	0-20	21-35	36-45	46+
3	0-30	31-45	46-55	56+
4	0-35	36-50	51-60	61+
5	0-45	46-60	61-70	71+
FMAA	Co Cdr / Ops O	Bn Cdr / AASF Cdr	SAAO	TAG

TOTAL RISK ASSESSMENT – NVG CREW CONFIG				
# CREW	LOW	MOD	HIGH	EXT HIGH
1	0-20	21-30	31-40	41+
2	0-25	26-35	36-45	46+
3	0-35	36-45	46-55	56+
4	0-40	41-50	51-60	61+
5	0-50	51-60	61-70	71+
FMAA	Co Cdr / Ops Offcr	Bn Cdr / AASF Cdr	SAAO	TAG

Low Risk - Requires CO CDR/AASF OPS Officer or see Note 1
*Mod Risk - Requires BN CDR/AASF CDR or see Note 1
**High Risk - Requires BDE CDR/SAAO approval
***Ex High Risk - Requires 1st General Officer/TAG

Note 1: In their absence, the next available whose assessed risk level is commensurate w/ their command level.

Note 2: One time flight for evacuation and MTF's conducted at night, combination of VFR, IFR, and VFR-on-Top will be approved only by the AASF or Company CO (or designated rep.) on a case by case basis.

There are a few main categories: duty hours, weather, crew experience level and additional factors.

There is one specific component I want to highlight: experience level and hours in particular aircraft. It takes 100 hours in aircraft before you drop two risk-assessment points. This is an acknowledgment that you don't know a lot the first 100 hours in ANY airframe.

The weather also plays a huge roll in the risk-assessment form. Notice how some conditions say "NA"? Our Army regulations prohibit us from flying in certain conditions. As a Part 91 pilot, you may not have these restrictions, but you could create these restrictions for yourself.

Spend some time with it and see what factors cause moderate and high-risk missions. Most of the time the missions will be low, but sometimes with bad weather and an inexperienced crew you can get into moderate risk quite quickly.

This is good education on what factors tend to be the most dangerous, like single-pilot operations, low experience level, fatigue and bad weather. Those are often "no-duhs," but when you see it written out on a risk-assessment form, it becomes more powerful.

I recommend you create your own and use it religiously until you get good at recognizing moderate-risk levels.

To give you some perspective, it's pretty rare for our flights to get to the moderate level. The mission has to be worth it to accept moderate risk. I have never been on a moderate-risk training flight. I can't think of a time when a training flight is worth moderate risk.

MILITARY TIP #18
ARE YOU ON THE SAME PAGE?

The crew brief is an integral part of every military mission.

Crew members are scattered doing different jobs when they prepare for a mission. Only one of the pilots may end up receiving the full mission brief from higher while the other three prepare the aircraft for the mission. This often leaves other crew members out of the loop.

A crew brief brings the crew together and gets them on the same page.

Don't think this only applies to the military! In my civilian job, the other pilot checks weather, NOTAMs and files the flight plan while I preflight. Sometimes, if something is up with the aircraft, I won't have time to go back and look at the weather and NOTAMs myself.

A quick crew brief is important to bring both pilots up to speed on the situation.

The military briefs can be quite lengthy, but yours doesn't have to be. Cover the important stuff:

- Route of flight
- Weather and NOTAMs
- Contingencies if encountering inadvertent Instrument Meteorological Conditions (if VFR)
- Emergency evacuation procedures and assign duties (e.g., who will use the fire extinguisher?)
- Specific crew duties (e.g., who will handle the radios?)

I have attached an example of a military crew brief. Go here to see the brief:

thinkaviation.net/fly-safer-and-smarter-resources

We do this brief prior to every helicopter flight. If I've flown with the crew before, I will abbreviate the crew duties, but if I'm flying with someone new, I will highlight specific things. For example, on instrument flights I will tell the pilot monitoring to handle the radio calls.

If you fly with the same person a lot, you can abbreviate it, but just be aware of complacency.

BONUS: The passenger brief is completely separate from the crew brief. It's also nonnegotiable. You should have both checklists laminated and available in the aircraft.

MILITARY TIP #19

DO WHAT YOUR COMMANDER TELLS YOU

Have you looked at the airman certification standards[14] since you got your license? I didn't think so. That doesn't happen in the military. We look at it a minimum every year.

We have a different name for it in the Army: the Commander's Task List.

The Commander's Task List (CTL) dictates how many task iterations each pilot must perform every six months. It also dictates numbers of hours we need every six months.

So, for example, I have to get 48 hours of flight time every six months. Five hours of it must be "hood" or "actual instrument" time, two hours must be at night and nine hours must be with Night Vision Goggles. The Commander may increase these

14 www.faa.gov/training_testing/testing/acs/media/private_airplane_acs.pdf

numbers if they feel their mission set needs a higher level of proficiency.

The semiannual period is an important distinction. It doesn't do you any good to do 10 iterations of a task all at once. But if you do 5 iterations every six months, you stay more proficient. The skill is fresher.

The whole point of a Commander's Task List is to maintain proficiency.

You should create your own CTL and run through it at least twice a year.

You should also create your own hour requirements as well (e.g., 40 hours every six months).

Don't think too hard about what your list should look like at first. Just use the airman certification standard and add to it based on your specific flying conditions.

Why do this?

You need to stay proficient, especially if you only get a check ride every two years. This list will ensure you stay not only current but proficient.

Note: You may have to fly with a CFI twice a year because some of these maneuvers you can't complete unless you have a CFI on board.

Before you think I'm going overboard by suggesting you fly with a CFI twice a year, I've said it once, and I'll say it a million times:

treat your aviation career like a profession, even if you don't get paid. Professionals spend time and money on continuation training.

If you can't afford an hour with an instructor pilot every six months, you don't have enough money to fly! Harsh, but true.

BONUS: Do you feel relieved you don't have to do stalls and engine failures every year? Watch out! That should be a huge red flag that you need to work on those procedures.

You should want to get proficient at those maneuvers. If you are avoiding them because they are uncomfortable, you have a lot of work to do. Emergency procedures must become second nature.

MILITARY TIP #20
DON'T BE LIKE THE MILITARY

This is the final military tip: don't act like you're a military pilot conducting military missions.

Why am I encouraging you to not act like the military after I just encouraged you 19 times to act like the military?

It's simple. I want you to exercise the most conservative approach when you fly.

What kind of civilian flight justifies pushing the limits? I would argue not much.

The military, by its nature, conducts dangerous missions. We often push the limits because people's lives are in danger. Oftentimes we say yes when we should say no. We don't like to leave a fellow soldier hanging if we can help them out, even if we don't feel completely comfortable with the flight.

I don't want you to act like we do!

Your passengers can check in to the nearest hotel if you can't pick them up.

Our military passengers might die if we miss a pickup.

Compare these to real repercussions, and you will see almost any scenario isn't worth it.

This is especially true of training missions. Just say no!

CONCLUSION

That's it! I hope you got at least a few good ideas from these military tips.

You can act like a pro, and I encourage you to do so even if you don't get paid.

Aviation is a serious business. You should treat it with respect. Part of that is developing habits of regular study, continuing education, and self-reflection.

I congratulate you for buying and reading books like this one, which will help you on your journey.

None of us is perfect. We have all made stupid mistakes, and I will continue to do so my entire career. The only difference is I have built up a series of habits to combat against the most common errors.

I also do regular training, and I have a daily study habit.

If you implement these same practices, you will notice a difference in your flying.

You will enjoy it more and you will become a safer pilot. Soon people will come to you looking for advice.

What the military does isn't rocket science. It's methodical and safety conscious. Because it's so methodical, it's easy to replicate.

GET YOUR RESOURCES

Before you go, I have created a page with links to all the resources listed in this book.

Go here to get your free resources:

thinkaviation.net/fly-safer-and-smarter-resources

ABOUT THE AUTHOR...

Sarah Fritts has been flying for more than 20 years. She is currently a pilot for a Part 135 on-demand operation and the Army National Guard. She flew OH-58D helicopters during the invasion of Iraq in 2003 and returned a second time with the 3-2 Stryker Brigade as an aviation liaison. Her final (and hopefully last) deployment was to Kandahar, Afghanistan, where she flew King Airs.

Sarah lives in the best city in the United States, Portland, Oregon, with her significant other and the cutest Schnoodle dog in the world.

She loves hearing from her readers, no matter what the platform. Pick your favorite:

: @thinkaviation | twitter.com/@thinkaviation

: ThinkAviationFacebook | www.facebook.com/thinkaviation

: ThinkAviationYouTube